Creek View Reflections

Inspiring Moments with an Oklahoma Poet

Barbara Zimmerman

Fresh Cup Publishing
Edmond, Oklahoma

Creek View Reflections: Inspiring Moments with an Oklahoma Poet

by Barbara Zimmerman, Edmond, Oklahoma

For permissions, contact freshcupbz20@gmail.com

Editing and Design by Darla Peake Hunter

Photos by Barbara Zimmerman

Cover design by Barbara Zimmerman and Darla Peake Hunter

Fresh Cup Publishing
Edmond, Oklahoma

ISBN 978-0-9904871-5-9

Dedication and Special Thanks

THIS BOOK OF POETRY is dedicated to my sister, Fran Anderson. Although there are eight years between us, our love for animals, flowers, and the beauty of God's creation has kept us close. Over the years, we've shared innumerable pictures of flowers, scenery, sunsets, and moon risings. Many of the poems and thoughts in this book were shared with Fran first. In fact, it was Fran who said, "You need to publish these poems. They are so peaceful, and people need peace in these days." Thank you, Fran, for being such an enthusiastic reader and supporter.

Above all, I give thanks to the Lord Jesus Christ for inspiring me to write. He gave me the eyes to see the beauty of His creation. Then, He enabled me to "paint" the pictures with precise words. May He be glorified!

A special thanks to Darla Hunter for your assistance in editing and design. You're such an encouragement and blessing to me.

Thank you, the reader, for reading my book. My prayer is that you will be drawn closer to Jesus Christ, the Lord and Creator of all, as you turn each page.

Contents

This Is My Father's World

This is my Father's world,
And to my listening ears
All nature sings, and round me rings
The music of the spheres.
This is my Father's world:
I rest me in the thought
Of rocks and trees, of skies and seas--
His hand the wonders wrought.

This is my Father's world:
The birds their carols raise,
The morning light, the lily white,
Declare their Maker's praise.
This is my Father's world:
He shines in all that's fair;
In the rustling grass I hear Him pass,
He speaks to me everywhere.

Maltbie D. Babcock 1901

Resting Place

I NEVER GROW WEARY
Of walking until I can rest:
Under the canopy of trees,
Refreshed by a gentle breeze,
Surrounded by bird and locusts
That fill my ears with sound.
Like the deer that seeks
And finds cool, refreshing streams,
Safe and hidden from danger,
To drink deeply
And be strengthened.
I will do the same.
Lord Jesus,
May my life reflect You,
As clearly as the trees' reflection
That rests on tranquil waters.
Lord Jesus,
May my life reflect your glory,
Like the full moon reflects
The brilliance of the sun.
May You be seen in me today.

Keeper of the Peace

WHENEVER I GO OUT-OF-DOORS,
No matter where I am,
Whether I'm sitting on my front porch,
Walking in the woods,
Sitting by a pond,
Or walking by a flowing stream,
May I always listen, observe,
And keep the peace.
May I always respect, honor,
And protect what God has created,
And may I teach and show others
How to do the same,
As a keeper of the peace.

Deep Unto Deep

THERE'S A DEEP PART of me
That draws strength
From lingering visits
In the deep woods.
Just being still,
Listening
To the gentle stream flow,
To the birds sing,
To the wind blow.
Receiving strength
And peace
From the Creator
Of the deep woods
And the Creator
Of the depths of me.

Still Flowing

STILLNESS
In the midst of the woods
Beside a flowing creek.
Rocks and pebbles
Try to block the way,
But waters ripple
As streams snake
Around and through
The rocky blockage.
New pathways form.
Peace
In the midst of the woods
Beside a flowing creek.

Among Trees

TREES ALWAYS SPEAK TO me.
They speak of strength, stability, deep roots,
Lifted hands, reaching arms, and endurance.
My Creator and Maker has been faithful
To surround me with trees all of my life.
For we were made to grow up among trees,
Both trees of wood and trees of flesh.
They provide shade in the heat,
Support during the storms,
Precious fruit to enjoy,
Seeds to sow, and seeds that grow.
I'm so thankful for the trees.
Thankful to be planted among them
By the Master Gardener.
Thankful to be with them
And to be one of them,
For trees always speak to me.

Place of Safety

ONCE AGAIN
My flourishing blue spruce
Proves to be a refuge.
Its bushy arms
Embrace new potted plants
Protecting from harsh winds.
Its thick foliage
Hides a nest in progress
While cardinal parents build.
Although my cat observes,
Her leash prevents her
From hunting up close.
Refuge.
Place of safety.
Hiding place.

The Cottonwood

THE GOLDEN COTTONWOOD
Stands tall above them all.
Its waxy leaves glimmer
With vivid reflection
Of the sun's light.
The giant with golden crown
Finally allows its leaves
To be released to the ground.
The wind carries
The yellow diamonds
Rustling and flickering
All the way down.
The dignified cottonwood
Stands tall above them all.
Branches lifted high.
Its leaves now color the ground
Settled in for fall.

Amazing Oak

TALL, FLOURISHING,
Green, growing,
Bending, swaying,
Unbroken.
Reaching higher, higher,
Extending farther, farther
Offering shade to many.
Stable, secure, dignified.
Deeply rooted,
Immovable.
Mature, fruitful,
Displaying the splendor
Of its Creator.
I am like you.
I am like you.

Gentle Awakening

NO DEER TRACKS IN the mud.
Creek is silent
With still reflections.
No flowing,
But small pools remain.
Even in the stillness,
Growth continues,
Light nourishes,
Waters satisfy.
Streams will flow again.
Now it's time to settle,
Receive,
Store up
For the days ahead.
Embrace the rays,
Feed on the light.
"Be still in knowing that
I Am God."

The Sighting

I STOPPED SEARCHING.
With the creek bed dry
And extreme temperatures,
My mind knew
They were safe and hidden.
But my heart still longed
To see them.
While outside this morning,
I spotted the doe
Headed out of the trees
Across an open field.
Two frolicking fawns
Followed close behind.
I held my breath and moved closer.
From one field to another
The doe sauntered
Looking for just the right spot
For the morning graze.
I remained hidden
Careful not to be seen
By this cautious Mother.
Two times she stopped.
Fawns ran close to her side.
All three looked straight at me.

I had to chuckle.
Their heads and ears formed
A delightful "Y" shape
Like saying, "Why, why, why?"
Mother found just the spot.
They feasted in peace
As she led them along
The fading boundary
Of an old wire fence.
At the bark of a neighbor's dog,
Eating ended, leaping began.
Over the fields they ran,
Graceful, white tails raised up.
Mama in the lead,
Babies close behind.
Finally, they reached
The safety of the trees,
And ambled back into hiding.

Tears of joy
Flowed down my face
They came to me.
They came to my place.

Time to Frolic

EARLY MORNING
Thunderstorms ended
Leaving saturated soil
And pools of water.
I walked
Down to the pond,
Singing with each step.
I stopped to watch
A deer frolic
Across the open field,
Its long white tail dancing
With each leap and prance.
Another deer
And two young fawns
Played as well.
Time to celebrate
The rain
And cool, refreshing breezes.
The dancing, prancing deer
Circled the whole field with glee.
Dancing, leaping, playing
Carefree, full of energy.
The foursome trotted
To a second grazing spot.

Celebration over,
Time to forage
Among the trees.
Camouflaged, hidden,
Unless you were looking,
In amazement, like me.

Be Quiet

IT WAS EIGHT O'CLOCK.
Mosie and I
Looked out the front window.
Deer check.
Strong winds blowing.
Early birds singing.
Squirrels chasing each other
In trees overhead.
Deer spotted.
"Be quiet, Mosie."
She sat as lookout,
I cracked the door.
Both of us froze.
There were eight deer,
Maybe more,
Eating grass and leaves.
They are still grazing
As I write.
I'm still amazed
At this stunning sight!

Time to Rest

THE FAMILY OF THREE
Moved across the open field.
First, the Mama
Strolled toward the shade
Of a group of trees.
Soon, her fawns
Appeared one at a time.
Walking, then running,
They joined Mama in the tender grass
Under the trees.
One fawn lay down
To rest while still eating.
Mama and other baby
Continued to graze.
Then, Mama lay down
In front of her babies.
Finally, the second baby
Meandered over by its sibling
And lay down in the grass.
All three rested in the shade,
Content from their meal
And enjoyed evening breezes.
Time to rest.

Hidden Treasure

IN THE WOODS
Looking back at me.
White tails flicking
And ears alert.
Mother stands still,
Three younger ones
Slowly pass by her
Into the deeper woods.
All is well,
So mother follows
Into the deeper woods.

Safe Place

I WAS HEADED OUT
For my morning walk,
When I saw a flick of white
In the shadowed woods nearby.
I froze.
Eye to eye
With a white-tailed deer.
We stared.
Her two fawns lazily grazed
Among the trees.
The doe joined them
For morning breakfast.
I watched,
Holding my breath
With delight and excitement.
Chosen once again.
I watched for twenty minutes
As they meandered
Through the woods.
They felt safe.
Then the wind picked up,
Leaves rustled,
And off they ran.
Two adults and another fawn
Had joined them.

They settled again
To finish grazing
On grass and leaves.
When they disappeared
Into the foliage
I took my walk.
As I returned home,
I startled two more deer
Who stopped, stared at me,
Then ran to the same
Forest area as the others.
Near to my home,
My safe place.

Stop and Gaze

GOING OUT MY FRONT door,
I paused when I saw
A deer
On the other side of the fence
In the woods.
I stood still and watched
As one doe and three fawns
Casually grazed
On leaves and grass.
Such a joy.
So dear.
Yes, I will stop and look
To see Your glory,
Your grand and dear creation.
Yes, I'll stop and gaze.
Thanks, Father.
I love it
When the deer
Come to me!

Snacking

ON THIS SUNNY, SPRING Sunday
I came outside
To my front porch
To enjoy my snack
And the cool breezes.
When I started munching,
I saw white tails moving
In the woods nearby.
Five deer ambled
And fed leisurely
On the green leaves
And foliage surrounding them.
They noticed me,
But didn't stop eating.
I'm not a threat
And I hope
They don't forget.
All I do
Is watch, enjoy,
And stay amazed
At their beauty, grace,
Alertness, and gentle ways.
All I do is
Hunt with my eyes.

Remembering the White-Tailed Deer

I FIRST SAW YOU eight years ago.
I lived down the street
And would walk to the edge
Of your land around dusk
To watch you graze.
When I had to move,
I moved on your land
Just across the street.
I'd look out my window
And catch you
Grazing in the open field.
Close to sunset I'd go
Searching for you
And usually found you
Grazing in another field
Down the road.

I'll never forget
When I discovered
Your hoof prints
In the soil near the woods
By my house.
Then I remember
Those delightful times
That I caught you

Feeding on the grass
In the woods nearby.
I'd spot you,
Stop and watch,
As still as a statue,
So I didn't disturb
Your peaceful ways.
I miss watching you
Run, leap fences,
Play and wag
Your stumpy brown tails
Lined with white.

Now, you've left Your land.
Too many new homes built
And you need room to live.
My heart is sad.
I miss seeing you
And find myself
Still searching for you.
I will always remember you.
And I will write about you,
So that others
Will remember, too.

The Deer

I MOVED TO THIS land because of you.
When I lived down the street,
I went for walks to the edge
Of your land, just to spot you
And watch you graze in the field.

Three years ago, I moved to your land
Because a friend recommended it
And I knew that you lived here.
For three years I've enjoyed finding you
Down by the creek
And even in the woods by my home.
I'd begin and end my day
Finding you nearby.
Your presence thrilled my heart,
Whether you're grazing, frolicking, running,
Playing, or just standing staring at me.
You settled me.

But now they've removed
Your forest refuge to build homes.
Your numbers have dwindled
Down to one or two.
Oh, how I've missed you.

I saw two of you this morning,
A pleasant surprise.
May you grow up and flourish
Somewhere else
Where there's space just for you.
Please forgive us for making you leave.

Leopard on the Sidewalk

IT WAS A BEAUTIFUL morning with the breezes just right and the songs of birds resounding from the trees. Mosie was hiding in the bushes along our home, and I was seated on the porch enjoying my morning tea. Suddenly, I heard something headed towards my apartment. Mosie slinked around to the front porch, ready to go inside.

"It's ok, Mosie. I'll check it out," I reassured her.

I got up from the porch and moved toward the side of my apartment.

Was it someone in their wheelchair or a friend strolling with her walker?

"Rickety-click, rickety-click, rickety-click." Closer and closer it came.

"It's a skateboarder!"

It wasn't just any skateboarder, but a young, blonde, little boy. He was barefoot and wearing a pair of boldly colored leopard shorts. He didn't notice anybody as he successfully rolled up and down the hill. He busily turned the corner and headed west.

What refreshing entertainment to begin a hot summer day. I guess it's time to expect the unexpected and enjoy free entertainment. My warning to those who walk on the sidewalk, though. Beware of the skateboarding leopard.

Stop, look, listen!

He could be on the move again.

Fireflies

FIREFLIES, FIREFLIES,
Within the woods so deep.
Fireflies, Fireflies,
Your light I'd like to keep.
Fireflies, Fireflies,
It's time I go to bed.
Fireflies, Fireflies,
I'll see you in my head.

Unnoticed

As I walked along the creek,
I was looking around me
And didn't notice
A mother Mallard
And her six ducklings.
She saw me
And swam nervously in circles
With her babies snuggling close.
I stood still.
The mother got her bearings
And paddled back upstream
With her babies clinging behind.
Special package,
Special delight,
Welcomed sight.

Taking a Pause

WHILE ON MY MORNING walk,
I couldn't resist taking a break
On the bridge overlooking the creek.
I put my walk on pause
To listen to the serenade
Of the Cardinals singing
From the trees above my head,
To feel the cool breeze
And listen to the Creek
Trickle over the rocks.
Even a small waterfall greeted me
With a gurgling flow.

I'm so glad that I paused,
For quiet reflection and
Time to breathe in deeply.
My Creator,
The One True God,
Lives in me
And surrounds me in this new day,
Even as His creation
Surrounds me with beauty
And sounds of life.
I will be still and know
He is God!

The Molting

I NOTICED YOU SLOWLY crawling
On the sidewalk
On a hot summer evening.
I knew you. I'd seen others like you,
But they were just empty brown shells,
The former home of a summer cicada.
I love your sound,
After you've changed.
I picked you up
And took you to a nearby tree.
It had rough bark
So that you could cling
And do your thing.
You climbed up the tree
Until you reached a short, dead twig.
You attached yourself to the twig
And seemed to be moving on the inside,
While slightly swaying on the outside,
Like change was fixing to burst out.
Adult cicadas were serenading
Your process of transformation.
I grew weary of looking up.
Darkness enveloped us.
I decided to go inside
To avoid being attacked by mosquitos.

Tomorrow when I check on you,
I expect to find your old home empty.
But I'm sure I'll hear you singing for joy,
Your change complete with wings to fly.
Thanks for crossing my path
And allowing me to help you
Reach your place of release.
I understand.
For I, too, am in process.

You Speak to Me

YOU SPEAK TO ME
On this hot summer morning.
Through cool breezes brushing against my face,
Through the sound of a woodpecker at work,
Through the sound of a gurgling stream,
Through the cacophony of locusts and Cardinals,
Through the coolness of the trees' shade,
Through a conversation with a veteran gardener.
You speak to me on this hot summer morning:
“Peace, be still.”
"Listen and enter into My rest.”

“I’m listening. I’m listening.”
“Yes, I will enter in.”

Singing Trees

THEY'RE HERE AGAIN.
Some people hate the sound they make,
Others, like myself, enjoy their sounds.
They make the trees sing like no other insect.
Their songs are so strong
That other sounds are simply not heard.
They are here again.
Some people hate the sound they make,
But I welcome their songs
And take time to soak in the cicada serenade.
Summer sounds, soothing songs.
Thankful for the singing trees of summer.

Tranquility

WHAT BEGAN AS A cloudy fall morning
With blustery winds,
Ended with cloudless skies
And slight breezes.
Twenty-eight Canadian geese
Grazed lazily in an open field
Before taking to flight
Into the settling sunset.
Two does and three fawns
Enjoyed eating in another open field.
They saw me and stared
Till I left them to eat in peace.
A great horned owl
Hooted "Goodnight."
A grey fox
Peeked out from a nearby bush,
Only to slither away
Into the darkened woods.
It's been a beautiful day
To be outside,
Enjoying the refreshing coolness
Combined with warm sunshine.
Autumn's here to stay.

Songs of Joy

RED-BREASTED MALE
House finch
Echoes songs of cheer,
"Babies will soon be here."
The female flits
Back and forth
From the nest
To the ground.
Both are busy
Preparing
For their young.
I simply enjoy
What's sung.
Three weeks passed
Three babies
Here at last.
New sounds heard.
Daddy still sings
While food he brings.
Mama speaks,
Babies peep.
Bobbing heads stretch
With mouths
Wide open.

Sounds of Peace

Whether it's the gentle gurgling
Of a winding stream,
Or the morning song
Of a bright red cardinal,
Or the slight rustling of leaves
Responding to a gentle breeze,
These are my sounds of peace.
Sounds that settle my soul,
Sounds that stir songs within.
Peace is the way to begin a day.
Peace strengthens for a new day.
Peace is the place to stay.
And Jesus is the only way.

Empty Nest

LOCUST SONGS
Are all I hear
As I look at the nest
That's all clear.
Gone
Are the songs
Of the protective
Male house finch.
Gone
Are the peeps
Of the chicks
Ready to eat.
Gone
Is the flurry
Of parents' wings
Fresh food to bring.
The nest is empty,
Empty of sound,
Empty of activity,
Empty.
But somewhere
Three young birds
Are learning the ropes
Of being grown up.

Such a treat
That this family
And I did meet.
And life grows on.

Signs of Change

THE SOUND OF LOCUSTS
Waned this morning.
I only heard them
As they were being
Chased and eaten
By birds.
What was once
A refuge for deer
Is now vacant.
Only shadows
Remain.

Autumn Evening

THE YELLOW GOLD SUNSET lingers
And drapes the western sky.
Three white-tailed deer
Race in circles
Through the wooded grassland
With tails lifted high.
The mother deer grazes
As her children play.
When the running
And leaping is over,
The fawns join their mother
For an evening meal.
Nighttime settles in.
Another beautiful fall day
Has come to an end.
Peace.
Stillness.
Rest.
Autumn at its best.

Listen

SCURRYING
On the rooftops
Click, click, click.
Squirrels
Gather acorns
For winter picnics.
Back and forth
From tree to tree,
In such a hurry,
It seems to me.
Cold winter ahead
Is what I hear them say.
Be prepared
Even today.
Up on the rooftops
Click, click, click.
Squirrels are preparing
For winter picnics.

Peddling on Pause

ON THIS COOL NOVEMBER evening
I had to go for a bike ride.
No wind and brilliant pink skies
Welcomed me.
Starting up the hill
I had to pause and admire
The bright, full moon rising.
In an open field
I noticed two fawns grazing.
I paused and wondered
"Where is your mama?"
Circling around the block
The pink sky screamed
And I paused
To just admire its beauty.
When I headed back home,
Twenty-five Canadian geese
Arose in flight as the sun set.
A doe appeared from the trees
And I had to pause to watch.
I tried to ride my bike
On this still November eve,
But I had to pause so many times...
I should've walked instead.

Good Night

BLUISH - PINK CLOUDS
Skirt the sunset horizon.
A fuzzy full moon
Hangs in the misty skies.
Robins make final statements
Before darkness
Swallows the light.
Still, very still.
Day ends.
Night begins.
Still, very still.

Rainy Autumn

SETTLED
At home
Surrounded by trees.
Lost in the scene.
With colors, sounds,
And rainy smells
Of the woods
On this chilly
Autumn day.

Heart Song

THE LAND SPEAKS OF winter.
My heart flows like spring.
Alive with thankfulness,
Vibrant with praise,
Living and moving,
Full and free.
Geese stream south
Sensing change in the air.
Leaves fall
Cushioning the ground.
Green grass
Turns shades of brown.
The land speaks of winter,
My heart flows like spring.

Candlelight Service

I LOVE CHRISTMAS EVE
Candlelight services.
Gathering together,
Family and friends.
Familiar songs.
Filling the air.
Familiar story
Of Jesus' birth.
Comforting.
Time to rest,
Be refreshed
By the joy
Of this event.
Songs sung,
Message shared,
Silent Night,
Candlelight.
My heart melts,
Tears flow.
I love Christmas Eve
Candlelight services.
Gathering together,
Family and friends.

Familiar songs.
Filling the air.
Familiar story
Of Jesus' birth.

On This Christmas Morn

ON THIS CHRISTMAS MORN:
I see the dawn breaking forth,
I feel the gentle breezes,
I hear the song of a cardinal from the woods,
I hear church bells ringing in the distance,
I see a duck in flight towards the pond,
I hear dogs barking from nearby neighborhoods.
On this Christmas morn:
I celebrate that the Light has come,
I celebrate that His Life flows,
I celebrate that His Love delivers,
I celebrate that His Peace rules.
I celebrate the One Who came,
Jesus, Creator and Lord of all!

A Song Was Born

A SONG WAS BORN
So long ago,
A Song carried by the Wind
From the Father's heart.

A star appeared,
Angels drew near,
Rejoicing,
Announcing His birth.

Wise men travelled
With gifts to share,
Worshiping,
Celebrating This One.

A Song was born
With piercing melody
To capture hearts
And set men free.

A Song was born
In the life of a Child
Who grew up
And lived among us.

He healed the sick,
Opened blind eyes,
Released His Song
And gave new life.

A Song was born
Within our hearts
Carried by the Wind
From the Father's heart.

Let's release His Song
With piercing melody.
Let the melody capture
And set men free.

Let's release His Song,
Let it flow
From the Father's heart
Everywhere we go!

New Beginnings

NORMAL SUNDAY MORNING.
Sunrise reflecting pink
Off the covering clouds,
Brilliant light show.
I dressed warmly
And went outside
In response
To its greeting.
No one
Was walking
Their dog,
Little movement
Of nearby traffic.
A flock of geese
Honked
As they came
In for a landing
On the pond.
Beautiful way
To begin
This brand-new day.
Not only a new day,
But a new year,
Opening up
Like a gift

Full of surprises,
Full of blessings.
Fresh new day,
Fresh new pathway.
Happy New Year!

Reaching Forward

IT TAKES COURAGE
To reach forward
Into the unknown,
Unseen,
Unexplored.
It takes courage
To reach forward
And catch new ideas,
New ways,
New perspectives.
It takes courage
To reach forward
And take the opportunity,
Take the challenge,
Take the advice.
It takes courage
To grow.

The Goose Parade

YES, CANADIAN GEESE HAVE a daily routine, and I'm an eyewitness of it all. I'm glad to say I live on a piece of land that includes a pond and open fields. As the sun is rising, the Canadians fly in for a landing on the pond for a brief swim and bathing time. About an hour later, they begin their parade.

Walking up the cement overflow ramp, they arise in single file. You'd think it was a military event, with equal distance between and marching in sync with each other. This morning, I counted 45 heads forming the parade. The leader leads them across the street to one of the open fields for a day of grazing. All day long they graze, and other geese may join them.

When evening approaches and the sun's thinking about calling it a day, individual geese begin to honk. They continue grazing with intermittent honks until the sun ends the day. The honking becomes more intense right before take-off.

They hit the skies in groups, heading for a resting place for the night. Many times, I've captured their flight silhouetted against the background of the oranges and pinks of our gorgeous Oklahoma sunsets. Of course, they're in formation for flight.

Their daily routine brings me daily delight.

First Day, First Snow

YOU COME AND YOU go,
You come and you go.
I've had my eye on you all day,
Oklahoma snow.
You snuck in with the cold
First day of the year, so bold!
Starting small, growing big,
Oklahoma snow.
Although I rarely see you
And you don't stay long,
You leave me with a happy heart song,
Oklahoma snow.

It's Snowing

IT'S SNOWING,
    Really snowing.
What started small
Now covers all.
Looking up
All I see
Is a flowing mass
Of white flakes
Now covering me.
It's cold, it's wet,
But I forget
I'm too busy
Enjoying the snow!!

Party Time

SNOW PARTY
At my bird feeder.
First the snowbirds,
Then several robins.
Soon the cardinals
And a house wren
Joined the gathering.
Then a boisterous blue jay
Announced His presence.
The party ended,
However,
When a squirrel decided
It was time to eat.

The Settling

THE STORM BROUGHT PEACE.
Few flakes, more flakes,
Blur of flakes blowing,
Yielding to the wind,
Covering everything.
Settled.
White skies,
White earth.
Settled.
Storms have the final say
At least today.
It's covered
And that settles it.
I'm safe inside
With windows to view
Thc predicted storm
On the news.
I'm settled.
The storm brought peace.
Few flakes, more flakes,
Blur of flakes blowing,
Yielding to the wind,
Covering everything.
Settled.

Frozen Skies

SPECKS OF SNOW
Blow sideways
Across the face
Of a red orange
Sunset display.

It's been a frozen day
With icy sleet-packed roads
And bitter temperatures.

It's been a frozen day
With little outdoor movement
Except for swaying branches
And busy, hungry birds
Swooping down to eat seeds.

It's been a frozen day
Just right for a puzzle,
Movie, popcorn, and nap.

The sun slips
Behind yellow grey clouds
Leaving pink streaks
Layered in blue.
It's been a beautiful frozen day.

Stormy Reminders

LAST SUMMER,
When I meticulously removed
Every bagworm I could reach,
To protect the stately evergreen
That beautified my front yard,
I didn't know.

This morning,
When I went outside in the bitter cold
To sprinkle birdseed generously
On the ground,
So that birds and squirrels could feed,
I didn't know.

I didn't realize
The evergreen would be
A vital safe place
For chickadees, robins,
Cardinals, and snowbirds
When the blizzard hit.

I didn't realize
The feeding area would be visited
By so many hungry creatures:

Birds, squirrels,
And even skunks and foxes,
When the blizzard hit.

I'm reminded,
What we protect and enjoy
In one season
Becomes a welcome refuge in another.
The seeds we are faithful to sow
Will feed more than we know!

Frozen Beauty

FROZEN BEAUTY,
Feathery branches,
Blanket of white,
Delightful sight.
Birds still sing,
Breezes blow,
Branches yield
Ever so slow.
Iced decor,
Waiting for more.
Time to embrace
Frozen beauty.
It's been a beautiful frozen day.

A Cardinal's Decree

ON THIS COLD WINTER day
I stood in the sun's warm rays.
Trees stood by,
Branches reaching into blue sky.
Then a song in the air
Caught my ear,
As a cardinal voiced
Its tune so clear.
I listened and heard
An encouraging word,
"Rejoice and sing,"
He seemed to bring.
My heart response
Didn't take long,
"Yes, I will rejoice
And continue my song."

In the Midst

BEFORE SUNRISE
I open my front door
To a pleasant surprise.
A concert from the trees
Was flowing on the breeze.
Due to the warmer weather,
My feathered friends
Refuse to wait and just don't care.
Spring sounds flood the air
In the midst of winter.
Robins, mourning doves, cardinals,
Chickadees, wrens
Release their songs
As the day begins,
Right in the midst of winter.

I lift my voice of praise,
For I agree
It's time to sing.
Let the praises ring,
Even in the midst of winter.
Singing praises to our Creator
Is just what we do,
Our voices we raise,
Right in the midst of winter.

Hope Arise

WINTER PASSING,
Snow disappearing.
Hope of a new season
Arises within.
Robins seen.
Cardinals sing.
Creation awakens,
Coming of spring.

A Winter Good-bye

I MUST SAY,
As I watch the snowstorm today,
"I will miss you, winter."
I'll miss
Beautiful snow accents
On the ground and all around.
I'll miss
Getting cozy warm
While harsh winds blow.
I'll miss
The stillness that snow brings
When busyness is put on pause.
I'll miss
Wearing warm coats and hats
That match the cold.
I'll miss
The wonderful season of Christmas
With family and friends nearby.

But I must say,
As white flakes keep on blowing,
"Let's get on with spring and stop this snowing."

Not Just Yet

I SEE NO LEAVES
On the branches of trees.
I hear no cricket
Or frog serenades from the creek.
I see no fish
Swimming in the waters.

But I do hear a cardinal song
And a robin's serenade.
But I do hear
The refreshing trickle
Of stream flowing over the rocks,
Fed by the melted snow.

The whole orchestra of spring
Is starting to warm up,
But not just yet.
The stage is set
And the sun is shining,
But the grand opening
Is not just yet.

First Responders

MORE DAYLIGHT,
More warmth,
More moisture,
Nature responds.
Green spurts up
From the ground
And bursts forth
From the branches.
Songs flourish.
Cardinals begin,
Robins join,
Others harmonize.
Colors appear.
Yellow daffodils,
Red crocuses,
Purple pansies.
Nature responds
To the morning light.
Nature responds
To the hope of spring.
I respond as well,
With a joyful sound
And a song of praise
To the Giver
of Life,

The Maker
of heaven
and earth.
"Let everything
That has breath
Praise the Lord!"

Nestled

As I looked out my window
At a large corner bush,
I noticed a splotch of red.
Looking closer,
I discovered a cardinal.
It wasn't singing,
It wasn't braving flight
On this very windy spring day.
It was nestling
Within the thick
Green leaves of the bush.
It stayed there
Long enough
For me to notice,
Long enough
For me
To write this poem.
Long enough
For me to settle.

Chosen

I'VE BEEN CHOSEN
By a radiant red cardinal.
Each morning
He greats me
Singing from
An oak tree
In my yard.
He's declaring
His territory
And I'm blessed
To be in it.
Morning and evening
And throughout the day,
His cheerful voice
Floods the air.
My heart rejoices
As I listen to his sound.
Spring is here.
New life bursting forth.
I will join and sing
For I've been chosen
By a cardinal,
And chosen
By the King.

Before the Storm

NO WIND.
Perfect stillness.
Cardinals and mockingbirds sing on
As if it were a sunny spring day,
Rather than a sullen, cloud-covered one.
Weathermen predict dangerous details
Preparing us for what lies ahead.
People hurry home to a safe place.
Highways flow with homebound workers.
Unusual way to begin a restful weekend.
Saturday and Sunday should be beautiful,
But today holds impending storms.
So, hurry home, stay safe,
While spring storms forth.

Smart Squirrel

THIS MORNING I AWAKENED to thundering clouds, lightning, and steady rain. It was dark and gloomy, so Mosie and I went to my office to watch the storm progress. Mosie took her place stretched out on the piano bench. I stood by the window that I had slightly cracked open.

I noticed that a small female squirrel hunkered at the base of the tree near my birdbath and bird feeder. She started towards her morning treat but scampered up the tree trunk as the storm progressed. She stopped where the large branches intersected, curled up, put her tail up over her back, faced my window, and closed her eyes.

During the downpour and lightning, she remained in a furry ball. When the rain let up, she stretched, exposed her tummy, and vigorously licked her fur. It reminded me of Mosie's daily cleaning process.

After a while, she zipped up the tree beyond what I could see. Now I know how squirrels can protect themselves during thunderstorms. They roll up into a squirrel "furball."

Overflowing

IT WAS A WONDERFUL spring day,
Covered with gloomy, grey clouds
Full of moisture.
Intermittent showers were
Accented with light rain drops.
Moisture, moisture in the air,
Water flowing everywhere.
Overflowing porches
And riverbeds,
Sidewalks and city streets.
It was a day of flowing,
Overflowing, and saturation.
Now, after sunset,
Croaking tree frogs
Release their sound,
Two great horned owls
Echo their hooting calls.
Settled and quiet.
Rain has passed.
Night has come.

Tumultuous Storm

THIS DAY BEGAN WITH a bang—
Thunder from the clouds,
Lightning touching the ground.
Booms, crackles, and rumbles
Filled the morning.
Mosie the cat had to leave
Her towering bed by the window
Due to a threat—
Lightning bolts nearby.
Harsh winds caused
Trees to sway and
Flower stems to bow.
Porch time didn't happen,
Because the thunderstorm did.
Afternoon calmed down.
The rain faucet turned off.
Winds became breezes.
Evening thunderstorms
Finally moved through.
Porch time happened.
While the creek gushed,
And the tree frogs serenaded,
Mosie got cozy on my blanketed lap.
Peaceful way to end a crazy day.

Flowing Once Again

TODAY THE CREEK IS full of water
From the storms yesterday.
Today a large, moss-coated turtle
Swims and feeds in the creek.
Today the stillness of the morning
Welcomes sunlight through the trees.
Today I've ventured
Back to my place of reflection and inspiration.
This is the day that the Lord has made.
I will rejoice and be glad in it!
Hope arises,
Fresh joy emerges,
My heart is encouraged,
For I have met with Him
In our secret place.
My River is full,
Flowing with His Life,
His peace,
His promises to me.
Flowing once again.

Emergence

THE NEARBY FOREST
Once draped
With browns and grays,
Is now transformed
Into brilliant shades of green.
The emergence of spring
Carries colorful proof
That life prevails.

Lost, but Found

SUNDAY,
Sunny spring day,
Fifteenth of May.
Seated in the shade
Facing the nearby woods.
Time to get lost.
Lost in the cool breezes,
Lost in the dancing shadows,
Lost in the sounds.
Singing male cardinal,
Chirping robin,
Cawing crow
All join in the show.
Lost in the greenery,
Lost in the growth,
Lost in the thick vines,
Lost in the towering trees,
Lost in the woods.
Found is a rested mind,
Found is renewed peace,
Lost, but found.

Heart Tuning

A NEW DAY IS dawning.
Cardinals declare it,
Robins and crows join in.
Two owls share their
Resounding hoots
And honking geese
Herald their flight
In perfect V-shape style.
The slight yellow
And pink horizon
Slowly becomes
Lighter, lighter, lighter
Until the sun peeks
Over the horizon,
Soon to appear
Full face.
My heart
Also rises
And tunes up
Its melody
Of praise
To the Creator
Of mornings
And new beginnings.

Beautiful Ending

THIS DAY HAS ENDED
With beautiful pink and
Blue rays of sunset,
Two deer grazing,
And a crescent moon settled
In a dusky blue sky.

Cool breezes remain
After a recent rain.
I stand amazed
As I see it all.
I stand amazed
Once again.
My heart rejoices,
Lightning bugs appear,
Two deer still grazing.
Amazing!
Light reflects in a pond nearby.
Beautiful ending to a beautiful day.

Discoveries

WHAT STARTED AS A morning walk
Became an adventure.
When I spotted the wild turkey
And the three deer
Moving through the tall grass,
I knew I was in for a treat.
Although the grass was wet
And the ground was muddy,
I found myself in the woods
By the flowing stream.
The morning shadows
And the gentle breezes
Blowing through the trees
Settled me, put me at ease.
Out in God's creation
The works of His hands
Gather together in song and dance.
This morning I joined them,
Celebrating with song,
Discovering beauty,
And delighting
In the Maker of it all.
Maker of me and
Maker of all I see.
Glory, Glory, Glory!

Made in the USA
Monee, IL
03 November 2026

67386982R00069